How Do You Start Keeping Bees?

Dr. Steven A. Josephsen

ARCHWAY
PUBLISHING

Archway Publishing books may be ordered
through booksellers or by contacting:

Archway Publishing
1663 Liberty Drive
Bloomington, IN 47403
www.archwaypublishing.com
1 (888) 242-5904

ISBN: 978-1-4808-8398-7 (sc)
ISBN: 978-1-4808-8397-0 (e)

Library of Congress Control Number: 2019916239

Print information available on the last page.

Archway Publishing rev. date: 10/21/2019

For the love of honey.

Contents

Forward

Have you ever thought about keeping bees but had no idea where or how to start? This is the book for you. May**bee** you have always been secretly wishing you had a couple million bees happily making buckets of honey for you. May**bee** you have never even considered such a thing, but all of a sudden, out of the blue, the thought pops into your mind: Keeping bees is a good idea! May**bee** you simply assumed you would never **bee** able to do it and tried to put the idea out of your mind... but still the thought lingers! If you are not sure about it but you are willing to give the idea a good chance, this is the book for you.

1. Make the Decision

So, at some point, may**bee** the thought has crossed your mind: "Keeping bees is a good idea." And quite honestly, that's about all that matters. May**bee** you are seven years old? May**bee** you are seventy years old? Age doesn't matter in beekeeping! If you like the idea of keeping bees and you don't mind the work that it will take and you are okay with **bee**ing stung from time to time in the process, you're good.

Some people start keeping bees because they have a noble ambition to save the earth. Great. Some people start keeping bees because they love bees. Fine. Some people start keeping bees because they love honey. Who doesn't? Some people start keeping bees because they want to make mead. Yummm. And some people start keeping bees because they want to make money. Good luck with that! It really doesn't matter to me *why* you want to start keeping bees, because if you succeed in beekeeping, it is good for you, your family, your friends, your neighbors, the bees, the flowers, the fruit, and may**bee** even the whole planet! If you fizzle out, on the other hand, life as we know it continues unaffected. Bottom line, it is nobody else's business why you keep bees. Your decision is all that matters. For you, it may **bee** a hobby, a pastime, a business, or an all-consuming passion. I don't know. But the first and most important thing is that you want to do it! That's good enough for me. And that's your starting place. Just because!

Okay, so we've covered "Why do you want to start?" But where do you start? How do you start? What do you need to get started? You probably don't even know what questions you need to ask. So, let me start by telling you my story.

I first got interested in bees as a youth. I remember seeing a movie about bees made by the Moody Institute of Science. It was fascinating, and they had lots of cool facts about bees. The movie showed the wonder of God's creation as revealed through the amazing life of bees. It made a big impression on me. I was hooked. The only problem was that I had no idea how to get started in beekeeping, so the notion, "Keeping bees is a good idea" sat idle in my mind for a little while. As a matter of fact, it sat idle for well over 50 years. It wasn't until I began getting gentle notices from the Social Security Administration that I soon would need to register for Medicare that I actually gave the thought about keeping bees some legs. And that happened quite by accident.

My wife teaches Suzuki Violin. She has students as young as 5 and as old as dirt. The husband of one of her adult students had gotten the notion "Keeping bees is a good idea." He went out and purchased all kinds of bee stuff: hive boxes, frames, bottom boards, supers, bee books, tools, suits, gloves, a smoker and everything. He purchased a bunch of bees and went to a "bee school" only to have his bees vanish! They call that absconding! Poof! Gone! No more bees! Frustrated, discouraged, and may**bee** a wee bit angry, he wanted out. So he gave everything he had amassed to me. Now, there I am with a car full of cool bee stuff, and that long-forgotten notion, "Keeping bees is a good idea" sprang into new life for me. I was determined to get some bees! It just so happened that the family of a different violin student my wife was teaching is a family of beekeepers. They have lots of bees. Hundreds of hives. Millions of bees. They raise bees for money as a family business, so I bought a hive of bees from them. I bought an entire colony. Just like that... I **bee**came a beekeeper. Beeks, we call ourselves.

Do you think you might want to **bee** a Beek too? Hopefully, this book will help you decide.

Why keep bees? Why did I do it? Did I need a reason? Do you? Does anyone? I don't really know. May**bee** I just hated to see the free bee stuff I was given go to waste? I like bees, and I like honey, and I like mead, and yes, I want to save the planet! So for me, it was a no-brainer. I would keep bees, just because...

I **bee**came a beekeeper because I made the decision to do so. Make up your mind. That's really all it takes. You make the decision.

It seems like the first thing everyone is going to

tell you is that you need a good reason to **bee**come a Beek. But I'm not everyone. All I am saying is that you need to make up your mind to do it. The first step is really simple. Decide to do it or not! Don't worry about what others think.

Deciding to **bee** a beekeeper is more of an important decision for the bees than it is for you, however. For you, it's may**bee** not such a big deal. For the bees, it is a matter of life and death. So if you make the decision to **bee**come a beekeeper, you need to **bee** willing and able to commit to learning how to **bee** a responsible beekeeper. Responsible beekeepers keep bees alive!

Do you remember asking your mom or dad if you could have a dog or cat? (Or a chicken or a horse or a goat or a snake or a lizard or a turtle or a fish...) Your parents cautioned you about how much work it was going to **bee** but you wanted it (them) anyway, right? Let me do that parental-type-cautioning thing for you now.

Are you going to check on your bees every few weeks to make sure they are ok? Are you going to learn everything you can about your bees so you will **bee** a good beekeeper? Are you going to feed them when they are hungry and take care of them when they're sick? Are you going to *not* get mad at them when they sting you? (And sting you they will!) Are you going to spend your hard-earned money to get all the things you will need to keep bees? And you still think, "Keeping bees is a good idea?"

All right. You can **bee** a beekeeper. You have

my permission. But just remember, this was **your** decision.

Oh, and as I mentioned, you are going to get stung. If you are allergic to bee stings, don't **bee** stupid! Find something else to do. Seriously, I can't **bee**lieve I have to come right out and say that. And don't go skydiving without a parachute and don't play with fire... Wait! Beekeepers do play with fire. Oh, well, you get the idea. **Bee** safe. If keeping bees is going to kill you, don't do it. I rescind my permission! You have been warned.

These are some of my favorite bee-in-a-flower pictures. I just used my cell phone to take these photos. Now that I have bees, my peach trees are producing LOTS more peaches! I like to draw, so I drew the face of a worker bee, a honeycomb design, and a worker bee.

2. Read A Book

Here's my advice for as good a starting place as any: Read a book about keeping bees. The first thing I did, even before I got my bees, was to read the book I found in the pile of bee stuff given to me: *Honey Bees and Beekeeping: A Year in the Life of an Apiary* by Keith S. Delaplane. It's a very good book. It was a good start.

I'm a teacher, so of course I'm going to tell you to go read a book. This is a good strategy, however— that if you want to do something that happens to **bee** a bit complicated, you need to learn what it is all about. That means the first thing you need to do if you want to keep bees is to learn about beekeeping. Do basic research. Beekeeping is not terribly hard to do, but it can **bee**come a bit complicated. They are living things, and if you take it upon yourself to keep them, you need to know how not to kill them via ignorance! Sadly, I am already guilty of this! Besides, bees are fascinating creatures and I'm determined to get the hang of it! Get a book and read. You'll see.

There are lots of good books out there, too. If you

visit an online bee group and ask what book to read for starters, you will likely see every bee book known to exist recommended. For some odd reason, I think the very first book you need to get is this book, but if this is your second book, I don't mind! Buying a book is a small investment, and you will need to invest both time and money in beekeeping if you want to succeed. But if you still aren't sure about it, you don't even need to spend a penny. Go to your local library and check out a beekeeping book... or two. The first book you read isn't all that important, because if you are really motivated to keep bees, the first book you read will not **bee** the only book you read!

Where does one find beekeeping books, you ask? I have seen beekeeping books at hardware stores, farm and ranch stores, feed stores, and of course, online. Most places that sell beekeeping equipment will also sell beekeeping books. They will have the most popular ones too, so don't **bee** too fussy. Just pick out whatever they have that seems to pique your interest. I'm sure you will find lots of good information there. As soon as I finished reading my first book, I went out and bought *Beekeeping for Dummies* by Howland Blackiston and read that one too. It is easy to read. It is thorough. It was extremely helpful to me and gave me enough confidence that I felt I could go out and do the things I read about in the book. If you start asking around, there are *lots* of good books Beeks will recommend. Some of the books are rather folksy and some of the books are rather academic. *A Year in the Life of an Apiary* by Keith S. Delaplane is more academic. *Beekeeping for Dummies* by Howland Blackiston is a good bit folksier!

Some of the books are going to advocate this and some advocate that, and that's the way it is in life. Right? There are lots of opinions about beekeeping. I've heard the expression, "Ask two beekeepers a question and you will get five answers." I think you're more likely to get a dozen answers, though. You will quickly discover that some people are all excited about natural beekeeping, and they think natural beekeeping is the way you should keep your bees too. Some people are all excited about the latest in bee science and innovative technology and think you should **bee** excited about fancy new gadgets too. Some people think innovative technology is terrible and think you should keep bees the way the old timers did! With so many different options and opinions about beekeeping, it is a good idea to do a lot of reading, listening, and thinking about it first. Do enough reading to find where you seem to fit in best with regard to beekeeping philosophy and practice.

Bottom line: Beekeeping has been going on for thousands of years. Any one technique for beekeeping is going to work under some circumstance and not going to work under other circumstance. So, if you shoot for somewhere in the middle, you will do fine for starters. Once you have been at it for a while, you will find your niche, and, likely, you will grow more and more opinionated about it! And that's the way it is in life, right? So go read a book... Or two.

The first bee book I read was given to me along with the rest of the bee "stuff" I was given at the start of my apiary adventure. I really liked this book. It is written by an entomologist, so it has an academic flavor. But it set things out very clearly for me. I have already read it through a couple of times, but here I am reading it again.

3. Take A Look

I'm sure you've heard the expression, "A picture is worth a thousand words." If that's true, the videos you can find on the web are worth a ton of books. Admittedly, some of them are! You can find a video on the web about every aspect imaginable when it comes to beekeeping. It really helps to watch several videos on a single topic. You get a chance to find out the core issues on any topic and you will **bee** able to detect the fringe issues. I have watched several videos several times because I just can't remember everything the first time. Remember, beekeepers are opinionated and have different philosophies about how to keep bees. Their various philosophies influence the techniques they use and recommend. You will quickly see all of that when you begin to watch the videos each beekeeper makes. You will **bee** able to see where some things overlap and everybody agrees. You will see where some things seem to wander off and everybody seems to have a different approach to some aspect of beekeeping. That just means that there are some universal principles in beekeeping and some other principles in beekeeping that are more open for personal exploration and experimentation.

There are videos about installing bees in your hive. That's a good place to start watching! It's exciting, fun, and just a tiny bit "scary." You can easily imagine yourself standing there with a box full of 5,000 bees trying to figure out how to get them into your brand spanking new beehive. Then you watch a video, and the Beek in the vid makes it look oh-so-easy to do! After you watch a few of these, you start to think, "I can do that!"

There are videos about inspecting your hive. The first time you look at the bees, all you see is a swarming mass of bugs. You have no idea what you are seeing. Are they happy? Are they angry and about to attack? Is everything going well, or are they in serious trouble? Do they even know you are there? What is normal, anyway? It takes time seeing normal bees to **bee** able to quickly spot something out of the ordinary! Videos about inspecting hives give you a shot at knowing how to interpret signs of trouble from the hummmmm drummmmm.

There are videos about controlling bee pests. This is where one actually does the beekeeping! Humans are not the only ones who want to get at that honey! There are lots of potential problems bees may face. Some of their troubles are as big as a bear and some of their troubles are microscopic. How do I keep my hives protected from bears? Will wild hogs bother my bees? How do I get rid of hive beetles? What on earth is a wax moth? Do my bees have varroa mites? Are my bees about to swarm? Do they have enough honey for the winter? How do I know the difference between American and European Foulbrood? There are videos about all of this and more!

There are videos about bee swarms and catching them. There are videos about building swarm traps designed to "catch" free bees. I promise you that the first time you attempt to catch a swarm of bees, you will have a good deal of adrenaline pumping! It is amazingly gratifying to see them accept the new home you have prepared for them. It gives you warm buzzies!

There are videos about building beehives. You don't have to **bee** a master craftsman or a woodshop wizard. If you can follow directions, measure correctly, and cut a straight line, you have what it takes to make your own bee boxes. There are plans you can print. All you need are basic woodworking tools and skills and the motivation to "do it yourself." Okay, and may**bee** you just don't want to spend the extra money it costs to buy a ready-made, pre-painted, fully assembled beehive. I play Scrooge in *The Christmas Carol*; I get that! Save your money. Build your own hives. Watch videos so you can see how it's done.

There are videos about melting wax and all the cool stuff you can make or do with beeswax.

There are videos about mistakes rookie Beeks make! These are painfully fun to watch.

There are videos about raising queens.

There are videos about splitting hives.

There are videos about collecting honey! Beeks call this "robbing" the hive.

There are videos about extracting honey. Don't drool on the monitor!

There are videos about making mead... Salute!

Just keep in mind that for every movie you've ever seen where you have first read the book, the book is better, right? So, for me, the videos help me to see what the books are talking about. For me, the book is foremost; the video is follow-up. Videos are a vast, mostly free resource it is a shame to waste. You will probably want to watch some of the videos more than once because you will see different things from the place where you were vs. the place where you are now. So by all means, watch videos!

Bear in mind that some of the videos are made by genuine research entomologists and warrant close attention. Those guys really know their stuff. Some of the videos are made by enthusiastic wannabees who may serve you best as good examples of bad examples! Just try to pay attention to any important details they have on film, whether intended or captured by sheer luck! Some people post videos about beekeeping, and they are just as inexperienced as you. I watched a video the other day about a guy who was trying to move one part of his hive. I think he was trying to collect honey. But he was going about it all wrong. A few months ago, I could have seen this video and not been able to spot all the mistakes he made or know how he could have—should have— avoided disaster! But... he dropped part of the hive.

Whoosh! He had thousands of angry bees expressing their displeasure at **bee**ing dropped. I couldn't stop laughing as he ran off screaming in dismay! So, if the guy in the video you watch goes by the name "Goofy" and he's got a cold one in one hand and a hive tool in the other hand... you may want to count that viewing experience as "entertainment" rather than "enlightenment." Know what I mean? Viewer **bee**ware!

Bottom line here, keep reading beekeeping books and watching beekeeping videos until you feel like, "I can do this!" Watching bee videos and reading beekeeping books will expose you to bee jargon, equipment, and many different techniques and philosophies on beekeeping. Try to keep an open mind, but it probably will not take very long for your own philosophy to develop.

Just don't take rash action until you have a very good sense of knowing where you will **bee** comfortable when it comes to beekeeping practice and philosophy. Go with that! The bees will correct you if you are wrong. And count on things going wrong. You *are* human, right? The school of hard knocks is always in session somewhere! Count your blessings... Learn from your mistakes! And that is the life lesson for today!

I keep trying to put a bunch of videos into my book, but for some reason, it's not working! So here is a picture of me wearing my bee hood and jacket! May**bee** if I puff some smoke from the page, you will get a sudden uncontrollable urge to go eat a slice of honey on toast?

4. Join the Club

Beekeeping clubs and associations are all over the place. Beeks are a friendly lot, and they will welcome you like a long-lost soul! Bees live everywhere except Antarctica, so I am just sure there will **bee** beekeepers nearby. Of course, in a place like Texas, "nearby" means something quite different from a place like Rhode Island. Use the internet to find a local club or beekeeping association. Check with your chamber of commerce or county extension agent. They know all the clubs and the apiaries in your area. Call the local sheriff's office or look online or in the phonebook under pest control and then ask about local bee-removal people. People who remove bees often enough keep bees they remove. They will know all about the local clubs and associations because they probably are founding members!

Beekeeping clubs are a treasury of bee lore. These are the people for whom beekeeping is a way of life. There will **bee** newbees like us and veteran Beeks in the club. You may not even **bee** required to "join" in order to attend their meetings. Just go and try to absorb all that you can. Ask silly questions. They've heard it all before. There will **bee** people there who will have kept bees longer than you have been alive. There will **bee** others there who are relatively new at it. You will fit right in!

You don't even have to find a room full of people to join a club. Today you can join beekeeping clubs or groups on social media. My local beekeeping association has a Facebook page. I joined it. Facebook thought I might **bee** interested in beekeeping, and soon ads from other beekeeping Facebook pages began to pop up into my feed. I joined another... and another... and another. May**bee** you will know when to stop; I sure don't. You might want to search out "Beekeeping for Beginners" groups. There are plenty of those from which to choose. Lots of people will **bee** asking the same basic questions you are pondering, so it is a very helpful starting place. Veteran Beeks love monitoring the beginner sites, and they simply ooze with bee lore. Sponge it up! Oh, and keep in mind some of the old-timers have a sense of humor. Try not to feel hurt if you ask a silly question and get a silly answer. They mean well. We all need to laugh at our own silliness, right? Loosen up. Laugh at yourself. They won't leave you hanging. You'll get your answer from among the family banter. Because that's what it is... family interacting together. Consider it an honor they feel free to poke you in the ribs a bit.

The important thing is that in a very short time, you will **bee** in direct contact with Beeks all over the world whom you can go to and drink from the wells of their knowledge! Get online and ask a question. You may get 25 answers! Don't let that frighten you, because you will still find those 25 answers are not really 25 different answers. They may just **bee** 4 or 5 different options that get reworded and repeated several times over. It gets pretty easy to decide which of the 4 or 5 answers best fit *your* circumstances. You do need to bear in mind, however, that not all of the people in any particular online group live in the same hemisphere. So issues related to specific seasons of the year are going to **bee** flipped if you are chatting with someone from "down-under" or "up-over," as the case may **bee**. In that line, timing is also going to vary considerably between what's going on in Vermont versus what is going on in Arizona. What's happening up North right now is not what is happening down South. The weather on the East Coast is always different from the West coast. Things that are good to know about beekeeping in England may not mean a thing in Nigeria or Australia or Argentina. So you might want to just stick to clubs that are local at first. It's up to you.

Did I mention that Beeks are opinionated? You ask a simple question and next thing you know, a spirited argument is born in your feed! Here's what I take away from all that: Bees are flexible and adaptable. There are many right ways of doing things within a pretty broad range of possible answers. So, sure, you get lots of "different" answers, but the thing that makes most sense to you will probably work in your circumstances. As odd as it may seem, both sides of a bee management argument are apt to **bee** right. Most likely, the bees won't care. They will just deal with whatever confronts them and make the best of that situation. One solution may very well **bee** just as good as another! Go with what makes best sense to your unique situation, and things will likely turn out right enough. Having an insider's opinion is always a big help, however. So join the club!

All you REALLY need is a complete hive, a veil, a smoker, and bees. The hive has a telescoping upper lid covered with flashing, an inner cover, one deep Langstroth body with 10 frames and a screened bottom board. You can also use an entrance reducer to control the size of the opening to the hive. My bees are Italian. They were too busy to pose for this picture, so they said, "Ciao!"

5. Find A Mentor

The most important thing about joining a club is the possibility of finding a mentor. A mentor is the single most valuable resource a Newbeek can have at their disposal. A mentor is a teacher, a coach, a repository of bee lore, a cheerleader, a wealth of wisdom, and a friend. A mentor will give you that one-on-one attention we all crave... and NEED!

Did I mention that Beeks are a friendly lot? It isn't very hard to find someone from your local group that will take you under their wing as a mentor. Find a couple of mentors! Get all the help you possibly can! The best mentors just can't even help **bee**ing mentors. It is a defining characteristic of their **bee**ing. They will take you to their apiary. They will come to see your bees. They will take you shopping for bee stuff. They probably have extra bee suits and veils and gloves just waiting for an eager apprentice like you! If you start attending your local beekeeping club meetings, it shouldn't take too long to find out who is the kind of teacher who is best for your learning style. Pay attention to who answers your questions and explains things the way you best understand. Who seems to click with you? The person who may**bee** knows the most information about bees may not **bee** the best mentor for you. We each have our own teaching and learning styles. A match is good. A mismatch is bad. Take all the time you need to find your best mentor match. You will **bee** glad you waited.

A typical solo veil comes with a cool safari hat, but some are attached like a hood to a jacket or full suit (upper left). Feeders come in a variety of shapes and can **bee** used differently, as your needs may vary. I have an entrance feeder (top center) that uses a standard mason jar for sugar syrup. The "lid" to the jar is a plastic form that fits just into the opening of a Langstroth hive. A top feeder (upper right) sits on top of a hive with the hive cover acting as a lid. Sugar syrup is poured into the two chambers and bees crawl up through the opening in the center and stand on the floating "docks" to drink their fill without drowning! The honey super (lower left) is made from cedar and does not need to **bee** painted to prevent it from rotting in the elements. The grid mesh (lower right) is a queen excluder. It is plastic and the openings are big enough for all bees except the queen to get through. Using a queen excluder prevents the queen from laying eggs beyond the barrier, so bees fill those sections with HONEY!

6. Go to School

Did I mention that beekeeping can get a bit complicated? At this point, it is obvious you are on the path
to beekeeping reality. Up until now, it has all been
some crazy notion you have in your head, "Keeping
bees is a good idea." Here is where theory meets
practice. May**bee** you have been able to sort everything out and hit the bee box running... may**bee** not.
It would **bee** a shame if you have gone to all this trouble and you still are missing some important pieces
of the puzzle. Bee school is an organized program
designed just for beginners like you! The course will
bee comprehensive and structured and thorough.

Remember, I'm a teacher, so of course I'm going
to tell you to go to school! The nice thing is that there
are lots of options. Your local club will most likely have
a beginner's course they teach each year. May**bee**
you will find your mentor by going to this bee school.
It's a good place to look!

May**bee** you don't have a schedule that works with
the local bee school timeframes. Look for bee school
options online. Online bee groups regularly offer

courses. You will find some that are free and some that have a course fee. I managed to find a free course, and I **bee**lieve it was very nicely done. Remember, I'm a teacher, so when it comes to "school," I have very high standards!

There are different levels of bee schools from beginner classes to advanced beekeeping. I have only taken that one course, so I have no basis for comparing it to anything else, but I thought it was well-designed and very much worth my time. They had articles to read, videos to watch, and even tests to take. They did advertise their own line of bee stuff to buy, but they were not distasteful about it! At least, if you sign up for a free bee school course, it's easy to get your money's worth! Scrooge would approve of that. Go to school.

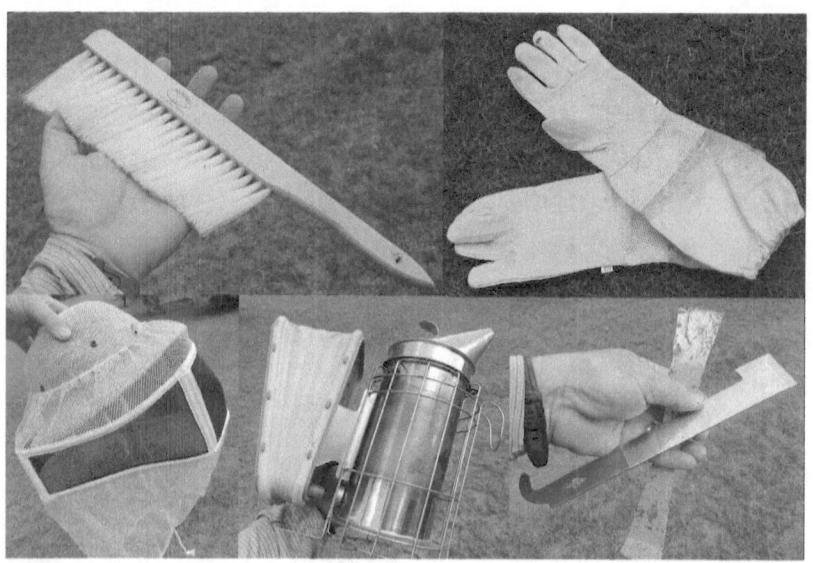

Beeks sometimes use a bee brush (upper left) to sweep bees out of the way without hurting them. I use goatskin gloves (upper right) because I just don't want to get stung on my hands. I have a spare veil in case a visitor wants to get up close and personal with me when I'm messing around with my "girls." The smoker (lower center) has a combustion chamber where you can put stuff that burns. It has a bellows to squeeze that puffs air into the combustion chamber to keep the embers glowing and puff smoke into your hive. Some people burn burlap. I use wood chips that I collect from under my wood-lathe. I have heard of people using dried, sweet-smelling herbs. You can find debates about what to put in there or just **bee** creative! These are my two hive tools (lower right). The old one has a bend on the end that acts like a crowbar for prying things apart. The new one has a J hook that works really nicely for pulling up frames after the bees glue them down. The new one is very sharp, and I already cut myself with it.

7. Find Your Hap-Bee Place

I know. This seems a wee bit obvious, but you need a place to put your hive. Here's what you need: The best place for your hive is out of the way a bit but still easy to get to. In the Northern Hemisphere, you basically want the opening of your hive facing any way but north. If you live in the Southern Hemisphere, you want your hive opening to face any way but south. You don't want a beehive right next to your front door, right? You don't want your beehive so far away that you feel put-out just to go there and have a little "look-see." Beeks LOVE going down and just watching bees do their thing!

A little bit of shade on your bee box is okay. All sunny all the time is okay and many Beeks prefer this type of location. Lots of shade is bad! Choose a place to put the hive box that has some reasonable shelter from the wind. If the bee box is out on a roof, a windy hilltop, or a high-rise balcony with strong crosscurrents, the bees might not **bee** able to get back in the hive without **bee**ing blown away! Some people build windscreens or put their bees inside a privacy fence.

If you have your hives in an enclosed garden, it is nice to give them enough open space that the returning workers can swoop down to the hive opening without stuff in the way. It doesn't need to **bee** very much room, but at least five or six feet of clearance helps them out.

Many people assume they don't have any place for bees, and they are wrong! Did I mention that bees live all over the planet? They live uptown, downtown, and across town. They live in the city, the suburbs, and the countryside. That means there are beekeepers that live from downtown to the boonies and everywhere in between! If you live in the city or a crowded neighborhood, take some time to look up "Urban Beekeeping." You might **bee** living next door to a Beek and never even know! They even make observation beehives you can install INSIDE your house, so you don't need to live on a farm with a thousand acres! You can keep bees just about anywhere! Ask your mentor or your club about where you should put your bees! EVERYONE will have an opinion!

At the very least, you need to **bee** able to stand next to your hive and have enough room to open it up, set the parts out of your way, and still have room for your own equipment when you do hive inspections or hiveway robbery. I use two 5-gallon plastic buckets to haul my bee stuff out to my hives. When I harvest honey, I need at least another clean bucket to hold my honey frames. People put bees in gardens, decks, or patios. They put them out in their back yard, out in a field, up on a roof, or out on a balcony. Wherever! Did I mention bees are flexible?

Bees hate, *hate* mowers and weed eaters! Anything you can do to minimize close encounters of the mowing sort will **bee** greatly appreciated (both ways!). I have my hives in a place where I don't need to mow nearby. They are in a natural meadow. One of my mentors says of lawn mowers, "You get one pass. Make it count." One of my mentors simply wears her whole bee suit when mowing. Take your pick.

There are some locations where beekeeping is regulated. You may live in a city, town, county, or housing development that has restrictions or ordinances you must follow. Your Homeowner's Association may not permit you to keep beehives, so do your legal homework before you set up your hives. Again, here's where local beekeepers can help keep you out of trouble!

Beeing a good neighbor about beekeeping goes a long way to help you keep out of trouble. Share some honey. Talk bee talk with them. Allay their fears. Some people disguise their beehives so that the unsuspecting busybody does not realize they are keeping bees on their property. Find a spot where your neighbors will not see the hives. Do what it takes to prevent neighborhood drama over your new hobby.

I picked this place for my bees because it is out of the way. It is in a meadow that I do not mow except for a pathway. It is mostly in the sun but shielded from the late afternoon glare, and if I get overheated, I can get into the shade just a few feet away. To begin with, I had deluded myself into thinking one hive would **bee** enough! Silly me!

8. Go Shopping

Have you ever seen an expectant mother prepare for the birth of her first baby? Oh, my goodness! They just about drive you crazy with all the preparation work that has to **bee** done before the baby arrives. Preparing to **bee** a Beek is not dissimilar! You really need to get all set up before your bees arrive. The first order of business is amassing a treasure trove of bee stuff! Here is what I think is the **rock-bottom minimum** of what you need to get started in beekeeping:

1. You need a complete hive.
2. You need a veil.
3. You need a smoker.

Keep in mind this is a minimalist list. There's more you may want. There's more you may need. Let's break this down more clearly for each of the above.

Beehives
There are lots of different types of beehives. Did I mention Beeks have different philosophies? Different philosophies about beekeeping are going to

bee evident in the choice of hive each Beek makes. People with a naturalist bent in their philosophy tend to choose a hive box that mimics the hive choices bees make in the wild. They may like the Warre hive or the Top-Bar hive. People who are more pragmatic or are most interested in harvesting the most honey possible may choose a Langstroth hive. People who are enamored by the wonders of modern technology may choose a Flow Hive. If you have decided keeping bees is a good idea, you read a book or three, watched hours of videos, joined a club, found a mentor, and went to school, you will by now at least have an *idea* which hive option is best for you.

Each hive type has its own "parts list." At a minimum, they have an exterior box and bars or frames that go inside for bees to use in making baby bees and storing pollen and honey. Hives can **bee** very simple homemade jobs that cost almost nothing. Hives can **bee** the latest in bee science and technology and **bee** very pricy! If you have a mentor, they probably have a favorite and will bend your ear telling you why it should **bee** your favorite. And they may **bee** right! A mentor can talk you through your selection decisions and can help you avoid **bee**ing ripped off.

I went with what I was given. I was given a Langstroth hive. This hive system has stood the test of time. The Langstroth hive makes up the vast majority of beehives you see. If you close your eyes and picture rows of white boxes lined up out in a field, that's what type of hive I'm talking about. The Langstroth hive and all of its components are completely interchangeable. You can make one from wood obtainable from any lumberyard and with only average woodworking tools and skill. You can print out the plans for building a Langstroth hive box from many websites. Most local stores that sell beekeeping equipment sell Langstroth hives and hive components. In fact, the only hives I have seen in local stores are the Langstroth hives. You have to go to an actual bee equipment store to find other hive boxes. Naturally, you can find Langstroth hives and components online. You can buy them unassembled and unpainted for less money than if they are assembled and painted. For a premium price, you can even buy them assembled and made from weather-resistant wood that never needs painting, like cedar.

A complete Langstroth hive usually has a bottom board, a brood box, an inner cover, and an outer cover. There can **bee** slight variations among these components. A starter hive will not come with any honey supers. Langstroth hive body components are all 19 7/8" long by 16 ¼" wide but they come in three depths: **deep**—9 5/8", **medium**—6 5/8", and **shallow**—5 3/4". The deep box is usually used on the bottom and is called a brood box because that is where the queen makes ba**bees**. You might hear a Beek talk about a "double-deep." This simply means the hive is **bee**ing given lots of room for making more bees by having two deep hive bodies at the bottom. The medium and shallow boxes are usually called "supers." They are mostly used for making honey! Supers get stacked up on top of brood boxes. A big hive may have half a dozen or more honey supers! You can buy a new complete hive kit for under $200 USD, and sometimes if you catch a good sale you can get them practically half-price! I have seen the exact same hive kit **bee**ing sold in two stores within 3 miles of each other, and one place was asking $100 USD MORE for it in store X than they were asking in store Z. Buyer **bee**ware!

The Langstroth hive has frames that bees use to build comb. The standard Langstroth hive has 10 frames, but they also come in an 8-frame version. Some beekeepers prefer the smaller 8-frame hives for a variety of reasons, the chief reason **bee**ing that they are lighter and easier to lift.

A frame is an internal structure that looks something like a cross between a picture-frame and a coat hanger. The frame is where all the action happens inside a beehive! Following the form printed into the foundation, bees will add wax to "draw out" the full honeycomb so that they can begin using those spaces for brood or stores. Standard frames are four pieces of wood with a sheet of plastic where the "picture" would go. Some frames are made of a single piece of molded plastic. The foundation within each frame is pressed with a honeycomb pattern the size bees normally use for raising workers or storing pollen or honey. Some Beeks prefer using wax foundations. Wax foundations can **bee** made "at home" with the right equipment or you can purchase them through a supplier. Some Beeks actually prefer to use frames without any foundations so the bees can make cells whatever size they want. Cells used for making drone bees (about 15-20% of hive cells) are a little larger than the normal cells. Cells used for making queens (less than 1% of all cells) are a LOT bigger than the others and look something like a peanut shell! Everything has its advantages and drawbacks.

Let's look at a Langstroth hive box from the bottom up. The bottom board of a beehive provides a foundation for the hive structure to rest upon. Typically, a bottom board will **bee** attached to the lowest box of the hive. As I mentioned, the bottom box of a hive is called the brood box. While there will **bee** some honey and pollen there, the main purpose for this portion of a hive is for raising baby bees. (Brood.) Between the brood box and the bottom board is the entrance of the hive.

If you look at the photo of the bottom board, you will see a wide board on one end. This is entrance of the hive and is called the landing board. When bees return home from foraging for nectar and pollen, they benefit from having a nice place to land. Returning bees may **bee** bearing a maximum load of nectar and pollen, and the wind might blow them off course. It's nice to have ample room to stray a bit and still make it safely "home." Some bottom boards are solid, and some have wooden slats. The openings in the bottom boards are for ventilation. Some bottom boards are fully screened to allow maximum ventilation. In climates where winter is frigid, ventilation is needed to allow moisture vapor to escape without frosting up the inside of the hive! While that's no problem for me in Texas, it may **bee** a BIG problem where you live! So far, I have only used fully screened bottom boards because of the Texas heat. In the summer of 2011, we had nearly 100 days in a row of 100-degree heat. I think my bees appreciate the screened bottoms with that kind of weather. You need to find what works best for the bees in your area. There are also specialty bottom boards that are designed to assist in pest control. These may cost more to begin with, but if they keep your bees alive, healthy, and pest free, you may find them "cheaper" in the long run.

The opening entrance of a Langstroth hive runs the full width of the hive box and is about 3/4" tall. If your hive is big and strong, the bees will appreciate having all this room to get in and out without running into each other. The big opening may **bee** critical for moisture control if you have a solid bottom board. If the hive colony is small, this could spell disaster, however. Humans are not the only things that like honey. Other critters and bugs will try to get into the hive, and so bees have guards for their entrance. They allow their sister bees to enter but keep out other intruders. If the entrance is too big for a small colony to defend, raiders can completely destroy the hive. The solution? Use an entrance reducer. The entrance reducer is a piece of wood that fits in the entrance and covers all but a portion of the access. Typically, there is a very small opening and a medium opening set at right angles to each other so that, depending on how you put in the block, only one opening is usable. Start with the smallest opening with a new colony of bees, and as the entrance grows overcrowded, switch out to the next larger opening or remove it entirely as needed.

The brood box is the heart of your hive colony. Most Beeks use a 10-frame deep body for their brood box. As the bee colony grows, many Beeks will add a second deep box (double-deep) for their brood chamber. Some Beeks only use a single deep for their brood chamber, and still others use only medium boxes (not quite as tall as a deep box) for brood chambers. The bees will not care. They will use the space you provide and the outside resources they find to grow their colony. Your job as a beekeeper is to help manage the space by adding or taking away boxes to meet their needs.

When a box is 75% - 80% full, it is time to put on an addition. If the population of the hive drops off for some reason (perhaps the colony swarmed and most of the bees that were there earlier are now gone) you may need to scale back the size of their habitat to 40% - 50% full. In other words, let's say you had a double deep (20 frames) set-up that was 80% full (16 frames full of bees). They swarm and now all you have left is 4 or 5 frames full of bees (15 empty frames). There is now too much empty space in the hive, and intruders can sneak into the hive and begin to cause trouble. To prevent such a thing from happening, take off the top deep (remove 10 frames). This will leave the remaining bees room to grow and repopulate the hive, but it does not leave more open space than they can manage. So add or remove boxes as needed.

The Langstroth hive has both an inner cover and an outer cover. The "telescoping" outer cover is designed to overlap all of the top edges of the hive box and keep out the elements. Since bees will try to glue this down, an inner cover is put in place first, followed by the outer cover. The layering of these two covers allows the beekeeper to easily lift off the outer cover and then use their hive tool to pry up the inner cover that the bees will have glued down. The inner cover is sometimes "notched" so that air can flow through. In a place like Texas, airflow is critical in keeping the temperature inside the box at a level the bees will tolerate. In colder climates, the airflow helps evacuate moisture vapor generated by the wintertime hive business operations. The outer cover is also typically finished off with metal flashing. Covers made with metal flashing will last year in and year out exposed to the weather all the while keeping rain, sleet, or snow out of the hive.

May**bee** you have seen advertisements or videos about the Flow-Hive that delivers honey in a jar with what seems like little more than a flick of a switch. I confess, I thought it was a complete scam for quite a long time! Fake news! Bogus! Magic tricks with the video camera! But I was wrong. There really is such a thing, and it actually works like they show in the videos. They are very high tech and very, very pricey! There are already cheap knock-offs that put the price down to merely twice as expensive as the Langstroth hive. But may**bee** you are into convenience or nifty technology? That's okay with me. The Beeks who have them love them. But many of the Beeks who don't have them loathe them... I say, don't get one if you can't afford one and don't want one. Remember, it's not what anyone else thinks that matters. What do you want? What is the price you are willing to pay for convenience? If a Flow-Hive suits you and you can afford it, go for it! The bees won't care!

If you get a beekeeping book and watch enough videos, they will talk about the various hives, and they will describe all the various parts. They will show you pictures and give you descriptions of each so you can make up your own mind. Bottom line, you need to go your own way in order to **bee** satisfied with your hive. Get the beehive that best suits your budget, your needs, and your philosophy of beekeeping.

Bee Veil

A bee veil is that funky netting thing that covers your head and neck. It keeps the bees off your face and out of your eyes. It prevents them from climbing down your neck and wandering around somewhere under your shirt. If you watch beekeeping videos, you will see some of the people tending bees in shorts and without any protective clothing whatsoever. On the beginner beekeeping sites, you often have someone ask, "Do I need to wear a veil?" The simple answer is "No." You don't have to wear a veil. You can tend bees without **bee**ing stung... right up to the moment when you do get a sting. May**bee** getting a sting on your face is okay with you. I've seen pictures of Beeks with nasty swollen lips and eyes and cheeks... because they didn't think they needed a veil... because they never needed one before. By the time you reach the conclusion "I need a veil," it's probably already too late to avoid **bee**ing stung in the face. Get a veil.

I have a couple of veils. I have one that fits over a cool safari hat and I have others that zip onto my jacket and full suit like a hoodie. I like the zip-on ones best because they stay put. The one on the safari hat is cool but seems to move around on my head while I'm tending my hives. I get distracted by that. Trust me. You don't want to **bee** distracted while you are messing around with bees. I have friends who wear them and love them. So who knows which type you will like best?

In some cases, a bee may see some interesting-looking hole in your head and decide, "May**bee** it is hollow in there? I should go check it out!" Does the idea of a bee crawling up your nose or into your ear sound like a great day out in the bee yard? Not to me, it doesn't! And my mentor said that happened to him! He was out working his hives and one crawled into his ear! I just can't even imagine having a bee crawling around inside my ear, licking up my earwax... playing pat-a-cake on my eardrum! No. Just no! You don't need a veil. But you surely do *want* to have one in case you ever accidentally drop something into an open hive... with 60,000 startled bees in there... Think about that. Take all the time you need. Get a good veil. You can thank me later.

Bee Smoker

Sending up a smoke screen is how Beeks can get away with all sorts of shenanigans inside a busy beehive. A bee smoker allows you to do this in a very easily controlled manner. The smoker lets you get some embers glowing and directs puffs of cool smoke into the area of your hive where you want to work. There are all sorts of ideas about how and why smoking bees works. I don't know which explanation is the best, and may**bee** there's some truth to all of it. I do know that it makes my job easier. You put flammable materials into the device and light it up, and you can make convenient puffs of smoke where and when you need them until your fuel is used up or the embers cool. You get to play with fire!

When you smoke bees, they make a sudden buzz and scoot away from that area. They immediately start eating honey. Bees that are eating honey are docile and not likely to sting. That means by use of the smoker:

1. You move lots of bees out of your way.
2. You get them to do something that calms them down.
3. You take their minds (may**bee** I should say noses?) off of whatever you may **bee** doing inside their hive.

Bees communicate via "smell," so having a puff of smoke is going to interrupt, mask, cover, and otherwise confuse other chemical smells that are going on inside the hive. So, imagine this... The bees are going about their business inside the dark quiet of a beehive, and all of a sudden, the blazing sun streams in! Some of the bees look up and see you messing around with their home and send out an alert hormone... but your puff of smoke overwhelms their alarm scent. So the rest of the bees don't get that alarm message. When the alerted bees send out their alarm but find that they don't get any confirmation of pending doom from the rest of their hive mates, they usually decide everything must **bee** okay. So the hive stays calm. As long as you work slowly, gently, and carefully, all you need is a few puffs of smoke from your smoker every once in a while to make your hive inspection go silky smooth.

Some Beeks don't use smoke. Some Beeks overuse smoke. I bet you could find at least 100 videos about how much or how little you should use a smoker. Use your best judgment and let your experience with your bees **bee** your guide! I recommend that you start by using a smoker. May**bee** as you grow in your bee craft, you will quit using smoke. May**bee** not. It's up to you.

Here is my very first harvest (upper left) of HONEY!
I used my food press (upper right) to extract honey.
Some people have used their hands to squeeze it out
of the comb. The most efficient way of extracting
honey is to use a frame spinner (not shown). Now
that's something to look forward to! Sometimes bees
make honeycomb in the strangest of places instead of
on the frames. We call that "burr comb" (lower right).
When burr comb is in the way, you have to scrape it
off. Here is some burr comb I had to scrape off (lower
left) that was full of honey. So I ate it! Yum!

9. More Stuff You Will Want

I know, I didn't tell you about getting bees. I won't forget that part, I promise. But before talking about installing your bees, there is more shopping you need to think about. My first list is the absolute minimum of what you need for beekeeping. There are some other REALLY handy things I want to tell you about that you will **bee** glad to own. I promise you that there is far more to buy than I will **bee** listing here. The next list of things I will tell you about is nearly... NEARLY essential.

More hives!

Why not double your fun? Set up two or three hives. If you think the crazy cat lady who has 25 cats is nuts, what about the beekeeper with 200 hives? This could **bee** you! It probably would take a while, but still it could **bee** you someday! Remember when I mentioned that at first it is hard for a beginner to know if something is wrong? If you have only one hive, you have no way of seeing odd things going on and having a sense of awareness that something is different between this hive and that hive. If you have several hives, you will **bee** able to spot differences between your hives! More importantly, when you see puzzling differences between your hives, you will know it is time to ask your mentor about what is going on.

Seriously, though, beekeeping is within the realm of animal husbandry. I have been keeping chickens for many years, and I have learned that within the circle of **life** is death. I know we don't like to use words like death or dead. We say, "He passed away." Everyone knows *he died* and is now dead, but nobody wants to say it plainly. Let me tell you plainly. If you keep bees, they will die! Sometimes this will **bee** your fault and sometimes it will **bee** bad luck and sometimes only the Lord above knows what happened! The only real art of beekeeping is to make sure that you consistently keep more bees alive than the ones that are dying. All the books happily tell you that a healthy queen can lay over 2000 eggs a day! But what they don't say is that it isn't unusual to also have 500 bees die a day. As long as the advantage goes toward good brood laying patterns (2000 − 500 = +1500), you are safe. But bees die. Queens die. Queens get old and quit laying 2000 eggs a day. If more bees are dying than are **bee**ing produced, the hive will die. And sometimes the end comes with dramatic swiftness.

In fact, lots of things kill bees. We had the coldest winter in decades when I got my first bees. Winter kills bees, but it isn't the only thing. Bees die from pesticides, disease, storms, flooding, drought, heat, parasites, animal attacks, old age, starvation, and even a neighborhood bully that decides to wreck your hives and kill your bees. A couple of teenage boys ransacked an apiary in Iowa, killing half a million bees and destroying tens of thousands of dollars' worth of honey and equipment in the process. As hard as you try to protect your bees and make life wonderful for them, some of them, perhaps many of them... even all of them might die. Ironically, "That's Life!" Having multiple beehives is Bee Life Insurance.

Honey Supers

If you raise bees, surely you intend to collect some honey. You can't take it all, because the bees will starve. After all, they make it because they need it for food. But they make more than they need, and you can take some. In order for bees to make honey, they need enough room to store it during times when blooms are abundant! Beeks call those times "honeyflow." A honey super is a hive component that can **bee** added to an existing hive. A honey super provides a nice place for your bees to make and store their honey during a honeyflow. A strong hive can make hundreds of pounds of honey. This is enough to fill several honey supers each year. Having more honey supers than you need is a nice problem. Having too few supers is a terrible problem that could doom your hive.

Hive Tool

You know that bees make honey, right? They also make stuff called propolis. Propolis is a sticky gooey tree-sap and wax concoction of gunk bees use to glue things down in their hive. If there is a crack, they will glue it together with propolis. When you put bees into a man-made hive, they find all the cracks in all the joints and seal them up nice and tight with propolis. So when you come along and try to pick up individual pieces of the hive, you will find them "gooed" together! The frames that are used in hives as the foundation for making honeycomb are loose. Bees will goo them together. The lid of the hive is loose. Bees will goo it down. Additional hive sections or honey supers stacked on top of each other are loose. Bees will goo them together.

A hive tool is really handy for dealing with this problem. It is designed to make hive inspection easier. The hive tool helps you pry them all apart gently. If you start pulling on something glued together with propolis, it will stay together for a bit until, POP! You break it loose. Bees find this experience strangely disconcerting (they hate it) and will fly out by the thousands to put a stop to the troublemaker! See where I'm going with that? If you have a hive tool, you can ease it into a glued joint, jiggle it in between the layers, and slowly, gently, and very carefully pry them loose without any sudden big snap, crackle, or pop!

Did I mention bees make propolis and smear it around on stuff? The hive tool has a putty-knife-like section that allows you to scrape off the goo. You can save the propolis and use it for several things. There are lots of videos about propolis. May**bee** you want to find a bee book and read about that... or just watch a video or two online!

Sometimes bees go a little crazy and make honeycomb in the wrong places, and the hive tool can **bee** used to scrape that off, too! Sure, you could use a putty knife, a paint scraper, a flat bar, and a screw driver to do all that stuff, but with a hive tool... you have it all in one handy piece. There are several varieties of hive tools, so you can start with a simple standard one for a few bucks and later treat yourself to something fancier. "What's that package that came in the mail from the bee store, Steve? You didn't go and buy more stuff for those bees, did you??" Shhhhh! Don't tell my wife I got more bee stuff!

Gloves

As with anything else, there are many arguments about wearing or not wearing gloves when working with bees. Let me tell you, I have bees that I can tend without gloves, and I have bees that I refuse to touch unless I have my gloves on! Here's the down-low on gloves: The books tell us that cow leather is not well liked by bees but that goatskin gloves are preferred. I have the goatskin gloves. Find a pair that fits and allows you to work in the hive without dropping stuff... (Remember, things that go bang inside an open hive are not popular with the winged ladies!) Some people like using those gloves you find in the hospital. I don't know why. They do absolutely nothing to protect you from a sting but I suppose they give you confidence and keep your fingers from getting sticky with honey or propolis. My wife HATES getting honey on her fingers! Some people like using those rubbery things people use for doing the dishes. Again, these gloves are not adequate protection from stings. Goatskin gloves are not 100% sting-proof, but they are very protective. (I have yet to **bee** stung through one, but I hear it is possible.) Bee gloves have gauntlets that keep bees from crawling in through the backside of the glove! The dishwashing kinds of gloves are completely open and you need to use a rubber band to keep bees from crawling in the open ends! Bottom line: bee gloves are... (wait for it)... very handy!

Bee Jacket/Hood

A step up from a simple veil is to have a bee jacket with a detachable veil-type hood. I have one of these and use it most of the time. It is my "go to" bee attire. The jacket is a tough, tightly woven material that stops most stingers from finding your skin. It will zip up and close all the little gaps a bee might find to get to your body inside. The fact that the hood/veil is integrated makes it easy to work around your hives without fear of your veil falling off or riding up at an inopportune time! They are hot to wear in the summer but tolerable.

Full Bee Suit

Take the idea of the integrated bee hoodie a step further and you have a full bee suit. If you have to deal with a very aggressive hive, you will **bee** happy to own a full suit. The only sting I got wearing one of those is when I forgot to zip up the bottom cuff and an overachiever bee found my ankle! The full suit can **bee** stuffy, so look for fully ventilated bee suits. If you are dealing with a bee removal that takes 10 hours and the temps outside are 100+ degrees Fahrenheit, you are going to need that ventilation!

Bee Brush

A bee brush is a soft-bristled brush that allows you to sweep a mass of bees aside without hurting them. You might want to brush them away from where you are about to place the lid of the hive. You might want to brush them off of a frame you are inspecting. You might want to brush a swarm of bees into your hive. A bee brush is also a very handy addition to your bee toolkit! Old Timers used to use a goose feather for a bee brush! If you are short on funds, head on down to a city park with a duck pond... Then play duck... duck... GOOSE! That ought to work!

Queen Excluder

If you are using a Langstroth hive, you can buy a queen excluder. This is a plastic or, in some cases, metal grid made to fit exactly on top of the hive. The grid openings are just large enough for worker bees and drones to crawl through, but not the queen. She will **bee** a smidge too big! Some Beeks put a queen excluder in between brood boxes and honey supers. This prevents any egg-laying activity in the honey super. With no eggs in the honey super, bees fill the honeycomb up there with honey!

If you capture a swarm, you might want to put a queen excluder between the bottom board of the hive (the exit is located there) and the lower brood box. This traps the queen inside the box and prevents the colony from re-swarming in case they are not 100% satisfied with the box you put them into. If Mom can't get out of the box, none of the rest of the bee family will leave her behind. They will stay put and start setting up house in the new digs. Some Beeks cut the plastic queen excluders into strips that can **bee** taped or stapled over the entrance. The beauty of this approach is that you don't have to move hive boxes to apply or remove it. You simply slap it over the entrance when you want and pull it off when you want! You don't want to leave the excluder over the entrance for too long, however. The first 2 or 3 days after capturing a swarm is enough to let them get settled and started in making the place their home. If you see bees entering the hive with pollen on their legs, the decision to stay has been made! Go ahead and remove the excluder. Of course, some Beeks hate using a queen excluder in any circumstance because it isn't natural. They want the bees to **bee** able to follow their natural instincts as much as possible with as little artificial intrusion as possible. This is just another one of those things you get to decide for yourself.

A Feeder

Bees eat honey. That's why they make it. But bees don't always have enough to eat. Sometimes you need to feed them to keep them alive! If you capture a swarm of bees, they don't have anything in their new hive to eat. Sometimes it is too hot or too cold or too rainy for bees to get nectar from flowers. If you order a bunch of bees from the mail, they don't have anything to eat. Making honey takes time and they will die if they can't make what they need fast enough. So if you want to make things easier for them, feed them sugar water. During the summer, you make the sugar water by mixing 1 cup of sugar with 1 cup of water. In the winter, you make it 2 cups of sugar to 1 cup of water. You need to heat up the water enough for all the sugar to melt but don't boil it! If you boil the sugar water, it will turn into rock candy once it cools, and you don't want that! But, hey, now you know how to make rock candy!

There are several different types of feeders on the market. Some feeders are simple to make for a DIY project. You can make a feeder from a 5-gallon bucket. If you read enough, you will see lots of options. You guessed it: you can find videos about how to make bee feeders! The important thing is to realize that there are circumstances when it is desirable, even necessary, to feed your bees. Bees live on pollen and honey. Remember, they collect sweet nectar from flowers and turn it into honey. If they do not have enough nectar, they can't make honey. If they don't have honey, they die. That means that as a Beek, you need to prevent death. Feed your bees!

There are two different approaches to feeding bees: internal feeding and external feeding or "open" feeding as it is often called. I have a picture of a Boardman feeder that can **bee** placed in the entrance of a hive. The nice thing about an entrance feeder is that they are easy to check on and refill. There are several different "top feeders" that fit on top of the hive. Some of these can **bee** serviced without opening the hive; others require you to open the hive. They all have openings down into the hive below where bees can come and drink their fill. There are feeders that can **bee** slipped into the hive in place of one of the frames. Frame feeders and some top feeders are hidden inside the hive. These are hard to keep track of, but the bees can keep other pests away from their food. Entrance feeders can sometimes attract greedy bees from other hives that will come rob the hive of all its honey, syrup, and pollen. That means death to the hive! **Bee** careful if you use entrance feeders. Open feeding is controversial. It is very convenient, but it carries risks. This might **bee** something to discuss with your mentor.

More and More and More...

If you go to a bee store or have them send you a catalogue, it is quickly obvious that I am only scratch-ing the surface here! Once you get STARTED keeping bees, you quickly start looking at all sorts of other great gobs of gleaming gizmos you might want. You can buy THOUSANDS of nifty items! You can get hive beetle traps, mite strips, oxalic acid vaporizers, pollen patties, and all sorts of accessories you never even knew existed. There are frame spinners that use centrifugal force to sling honey out of the comb and buckets with nozzles that help you store and bottle your liquid treasure. There are cutting dies that let you cut full honeycomb into pieces that fit perfectly into a honey jar so you can sell your honey with comb in it. They have cold capping knives and hot capping knives that you use to cut open the honeycomb once the bees have put a cap on it. There are kits for collecting and cleaning up beeswax. They make a roller that lets you make your own foundation forms from recycled beeswax. There are kits for grafting eggs into queen cups so you can raise and sell queens! I got a frame grabber. May**bee** you want one too? There are things related to hive management and protection. And there are all sorts of options for protective clothing and extracting and bottling honey. It is amazing!

If that's not enough for you, there are all sorts of cute honeybee kitsch items to buy as well. Everybody needs honeybee potholders or throw pillows, right? And how about a cute bee on the end of a stick you poke into your potted plants? Posters. Honey pots. Jewelry. Coffee mugs. Just you wait and see! Hide your credit card and go have a look.

I built two five-frame Nucleus hives (NUCs). If you are splitting a hive and now have a starter colony or two, put them in a mini-hive box! If you capture a small swarm, you don't want to put them in a box too big for them to manage. Use a NUC. Beeks often sell NUC hives with either 4 or 5 frames. This is a great way to get started keeping bees.

10. Buy the Bees

I know I am flirting with the obvious, but you are going to need some bees, right? There are thousands of strains of bees, but only a few have been successfully used for apiculture (beekeeping). Generally, honeybees are divided into two large groups, European honeybees (EHB) and the African honeybees (AHB). I bet you already know which of those to avoid! Don't worry, nobody sells AHBs! But within the EHB camp, there are several popular varieties from which to choose. There are Russian bees, Carolinian bees, Buckfast bees, Caucasian bees, and Italian bees. Italian bees are the most popular by far. Take some time to study each variety and find which ones seem to match your goals and temperament. Some bees are better in one region of a country but may**bee** not so well suited for where you live, so do your homework before you order a bunch of bees that are not going to flourish in your apiary! And for Heaven's sake, get help from your mentor when buying bees!!

My first bees were Italian Hygienic bees. That means that they have an inbred tendency to clean out more of the mites that can destroy a hive. Lord only knows what kind of bees I have captured from swarms. Bees that are captured from swarms are called feral bees. Feral bees are like a box of chocolates. You never know what you're going to get. One of my feral swarm catches has a bit of an attitude. At some point, I will likely get rid of (kill) the queen and give them a new queen (requeening) with a calm disposition. So far, if I am very careful, they don't get too upset with me messing around in their hive. We'll see.

Meanwhile, you can get your bees from several sources. You may **bee** able to get bees from a local Beek. You can order bees through the mail and have them delivered to your local post office. (They probably will not bring them directly to your house. You will most likely need to go to the post office to pick them up.) Package bees come with about three pounds of bees in a ventilated package that has an integrated feeder. There are videos about this. Watch and see how to go about "installing a package of bees." A package of bees can **bee** as little as $50 USD but averages more like $75 USD. I just saw an ad for a 3-pound package for $199 USD! So take your time and shop around.

In addition to the packaged worker bees, there will **bee** a separate package containing a mated queen with a few attendant workers. You can pay a bit extra to have your queen "marked." That means she will have a bright spot of paint on her back that makes her much easier to find during hive inspections. Once you place your bees into a hive box, the queen will need to **bee** added. There are some very specific directions for adding a queen that will **bee** discussed in the books I said you should read or in the videos I said you should watch. Failure to make proper introductions of a queen into a hive can end in apiary disaster. (The workers might reject and kill the queen.) So make sure you know how to add the queen before your package of bees arrive. There are several tactics to add a queen. You must find the method that seems most comfortable for you to employ. So, you are going to need a queen bee with a supply of workers as the barest minimum to start a bee colony.

Timing is everything, or so they say. So it is with getting started keeping bees. Early spring is the best time to start. The later in the growing season you start, the lower will **bee** your chances for successfully making it through the winter. So when did I start? At the end of summer! Silly me! But you may remember that I had to do a lot of things right to ensure my colony made it successfully. So do as I say, not as I did. Hum, where have I heard that before?

You can also start a beehive with a nucleus colony (NUC for short). This is a growing colony of bees with their own "home." A NUC will not only have a lot more bees, it will have resources! This means you get a queen and from 3 – 5 actual frames filled with drawn comb, brood, pollen, honey and the bees that established the colony. This will kick-start your new beehive and greatly increases their chances for success. Again, you may get a mated queen with attendants that will need to **bee** added to the hive or you may get the original queen. There are a lot of possibilities about how people sell NUCs, so you need to make sure you know exactly what you are getting. If you buy a NUC from a local apiary, they probably will have an established mated queen included. **Bee** sure to ask about what you are getting and what you may need to do for the bees. Many Beeks sell NUCs. They are going to go from around $130 - $180 USD, and the quality of the bees and queen and frames you get can vary widely. So sometimes paying less can lead to paying much more for the novice! Remember, get help from your mentor on this part! They can usually know if you are getting a good deal or not! Of course, not all bee problems are noticeable to the naked eye. An expert sure is a better judge of quality than a new**bee**!

You can buy a complete beehive. This typically is a box of 7 to 10 frames of bees and resources, and the queen will already **bee** present and busily doing her thing. All the bees will **bee** her children, so to speak. Getting started with a complete hive is almost like cheating because it is so easy. You buy the hive and take it home... done. Buying a complete hive can cost about $200 USD or more depending on circumstances. If you happen to catch someone getting out of the business, you can get amazing bargains. Many local Beeks will have an extra hive or two they would **bee** willing to sell. By the way, that's what I did! The purists will tell you to buy a package of bees, and they will extol the virtues of "beekism" you will gain by doing it the hard way. Well, it's not that hard, but don't spoil the flow here! But you should **bee** aware that you have a choice of starting from scratch vs. starting with a mile-long head start. In the end, nobody will care if you got started the easy way or the hard way!

I got my first bees at the end of the season, and I knew there was not much time for my bees to get ready for winter. If I had purchased a package of bees at that time of the year, they would not have survived. A NUC (nucleus hive) had only scant chance of survival. So it made perfect sense for me to start with a full colony. As chance would have it, we had the coldest winter in decades. It is a good thing that I did a bunch of things right that made it possible for my bees to make it through the winter without trouble! I listened to my mentor. I read my books. I watched hours of videos. I took a beginner's bee school course. I made the best decisions I knew to make (not without trepidation!). And my bees survived! If I can do it, I bet you can do it too!

The way bee populations spread in nature is through a biological process known as swarming. The colony will split itself into two groups. The old queen will fly off with more than half the colony of bees and attempt to find a new hive location and set up shop there. The remaining bees of the original colony will have a new queen in the making before the first half of the colony departs. So within a couple of days of the swarm leaving the old hive, a new queen will emerge from her cell. She will mature a bit, and in about two weeks she takes off for a mating flight. Upon her return, she takes over the old digs, and the original hive goes on as before. If the original swarm is successful, there will now **bee** two hives of bees where once there was one. Typically, swarming happens when a hive is strong and healthy, and it is the middle of spring when flowers are everywhere to **bee** found. Once the weather gets hot and the flowering plants slow down to a summer crawl, bees do not swarm so much.

If you are observant, you can detect when your hive is preparing to swarm, and you can beat them at their own trick. You can create an artificial swarm. You can split your hive! There are several methods for doing this that you can read about in your personal study. The basic idea is that you take one strong hive, divide up the resources, and separate everything. You end up with one queen in one half of the colony and no queen in the other. You can order a mated queen for this other box of bees and integrate her into the queenless half, or you can order a queen cell (a developing queen about to hatch) and put her into the queenless half. The bees will smell that it is a queen and they will adopt her as their new mother when she chews her way out of her queen cell. On the other hand, you can simply separate the two half hives, and if you made sure that there were fresh eggs in the queenless half, the nurse bees there will select a couple of them and turn them into queens. Either way, at the end of the process, you will have two hive colonies instead of one! There are lots of articles and videos that will teach you the fine points for splitting a hive successfully.

There is one more way to get bees that I wanted to save for last. This is for you die-hards who have to do things the really hard way. You can capture a swarm of bees for free!

As I said earlier, a hive of bees will swarm when the conditions are right. The swarm can contain tens of thousands of bees and a queen. They will climb out of the old hive and fly off to a nearby resting place. They are not very particular about where they will stop. They may rest near the top of a tree, or on the bumper of a car, or on a branch of a bush or on a post of a fence. Google "bee swarm photos" and you will see them in the craziest places! There is no telling where they may end up. Wherever they land, they will clump up together and form a ball of bees with Mom at the center. The cluster may hang there for an hour or two. They may hang there for several days. But for as long as they are there, the swarm is sending out scout bees that are only looking for a new cavity the swarm can use to make into their home.

When most people spot a swarm of bees, they freak out and assume they are "KILLER BEES" (AHBs), and often they will attempt to strike first! They will spray them and kill them! They may call a pest control service that will also simply kill them. Don't let that happen! If you cannot give these bees a new home, call your local beekeeping group and someone there will **bee** delighted to come scoop them up!

On the other hand, now you could **bee** the one

to do it! If you find a swarm of bees, you can catch them and dump them into your empty hive box. The bees will look over your box and most likely decide to stay. If they are hanging on to a branch, carefully cut the branch, hold it over your open beehive box, and give it a good hard shake! They will fall down into the box. If they are hanging somewhere, you can use a cardboard box or plastic bucket or even a trash-bag to catch them and then empty them into the beehive. They will quickly realize "this is a great spot for us to live," and you can take them home with you! The first swarm of bees I captured was so simple; it took less than 10 minutes. I captured one swarm that took much longer because I couldn't reach all of the bees. I did capture the queen right away and as soon as she was dropped into the hive box I had for them to use, the rest of the bees sniffed out her location and came to where she was. That took about two hours, but eventually they all figured out where Mom went and joined her in her new home. You can find out about bee swarms from your local bee club! And tell everyone you know that you capture swarms, and you might get lucky. There are lots of videos about catching swarms, but I can tell you that there is great satisfaction in catching one all by yourself!

This was a large swarm of bees that I captured! It was my second attempt at capturing a swarm, and I admit that I was feeling a bit intimidated. I took a guess that the queen was right in the middle of this side of the swarm, and I was right! I climbed up the ladder and made one sweeping brush stroke to knock the bees into a 5-gallon bucket. I quickly climbed down and dumped them into a hive box I brought along as their new home. Once Mom went into the hive box, the rest were glad to follow!

Let's Sum It Up

Getting everything you need depends on your skill as a bargain hunter, as a handyman, and with a bit o' luck. You can make most of this stuff yourself for nothing more than the cost of the materials and your time. You can buy used equipment at bargain prices. Some people may give you things you need, but if you had to buy everything, it is going to cost at the minimum between $200 - $500 USD to get started. You can order everything—and I mean EVERYTHING—online. There are some actual bee stores you can visit, too, but not very many. The nearest actual store to me is Mann Lake, and they are about 100 miles away. Bee supply companies sell basic starter kits that have the sort of things I have been telling you about. They even offer deluxe starter kits that can go up over $1000 USD. So, it's up to you. WARNING: Beekeeping can **bee** an obsession, so there's no end of things to buy! Don't tell my wife, but I've already spent nearly $1000 on bee stuff. But... But... I needed it!

Here is a very brief list (my apologies to the

THOUSANDS of others that didn't fit) of some handy North American bee resource outlets:

Atwoods Ranch & Home www.atwoods.com

Bastin Honey Bee Farm www.bastinhoneybeefarm.com

Dadant & Sons www.dadant.com

GloryBee www.glorybee.com

Kelly Beekeeping www.kellybees.com

Mann Lake www.mannlakeltd.com

Miller Bee Supply www.millerbeesupply.com

The Bee Place www.thebeeplace.com

So... did you decide keeping bees is a good idea? If so, here's your "To-do" list:

- Read a bee book or three
- Watch bee videos
- Join a local beekeeping club or association
- Join social-media-based beekeeping interest groups
- Find a mentor
- Go to a bee school
- Find a place for your hives

Go shopping:

- Complete Hive (or two)
- Veil
- Gloves
- Jacket/hood
- Suit
- Smoker
- Hive Tool
- Honey Super
- Queen Excluder
- Bee brush
- Bee feeder
- Bees

Voila! You are a Beek!

Good Luck!

About the Author

Dr. Steven A. Josephsen (pronounced "Joseph sen"), AKA Dr. J., grew up hating school because as a divergent thinker, he was always seeing things differ-

ently from the way his teachers thought he was "supposed" to see them. After high school he had no idea what he should **bee**come, but the example of an excellent teacher in college showed him what an awesome experience teaching and learning could **bee**, so he **bee**came a teacher. He taught 2nd grade for two years and then developed a program for the Gifted & Talented. The program spanned the grades 2 – 7 and covered all subjects. He served as the director/teacher for this program for the next 16 years. In his "spare" time he taught science classes before moving into Higher Education. He had the privilege of raising three children, one girl and two boys. They are all grown and married now. He's a grandpop now, too.

Dr. Josephsen did graduate work in the areas of Instruction, Gifted Education, & Educational Technologies. He has been an Assistant Professor at Stephen F. State University since 1996.

Once the "nest" was empty, Dr. Josephsen and his wife moved "out into the country" where he raises pine trees, geese, ducks, chickens, and now bees. He enjoys just about anything you can do outdoors as well as singing, playing hand-percussion instruments, photography, drawing, painting, woodworking and making wooden bowls on his lathe. He also enjoys growing things, but, to his utter consternation, considers himself a serial orchid murderer! He cannot help himself in this addiction. He sees beautiful orchids and buys and takes them home, where he slowly tortures the poor things to death. He has tried every suggestion from successful orchid growers to no avail. His only wish is, "Don't cry for me phalaenopsis!"